Sir Cumference

and the

Dragon of Pi

A Math Adventure

by Cindy Neuschwander
illustrated by Wayne Geehan

Charlesbridge

For Bruce with love and with special thanks to Elena — C.N.

For my wife Susan and children Greg, Jonathan, Douglas, and Amy — W.G.

Published by Charlesbridge • 85 Main Street, Watertown, MA 02472
(617) 926-0329 • www.charlesbridge.com

Printed January 2012 by Sung In Printing in Gunpo-Si, Kyonggi-Do, Korea
(hc) 10 9 8 7 6 5 4
(sc) 25 24 23 22 21 20

Library of Congress Cataloging-in-Publication Data
Neuschwander, Cindy.
      Sir Cumference and the dragon of pi: a math adventure /
by Cindy Neuschwander; illustrated by Wayne Geehan.
         p.   cm.
      Summary: When Sir Cumference drinks a potion that turns him into a
dragon, his son Radius searches for the magic number known as pi which
will restore him to his former shape.
      ISBN-13: 978-1-57091-166-8; ISBN-10: 1-57091-166-5 (reinforced for library use)
      ISBN-13: 978-1-57091-164-4; ISBN-10: 1-57091-164-9 (softcover)
      1. Circle — Juvenile literature. [1. Circle. 2. Pi.]
I. Geehan, Wayne, ill. II. Title.
QA484.N48 1999
516'.15 — dc21   99-10185
                                                      CIP

Sir Cumference and his son, Radius, sat in the shade of a tree, enjoying a delicious midday meal. Partway through the lunch, Sir Cumference grabbed his stomach and doubled over in pain.

"Ooooh, my belly!" he wailed. "It feels like fire! Radius! Run to the castle. Find the good doctor and get me a cure."

Radius ran to the castle
and up a winding staircase to the
doctor's workroom. The doctor was out. Radius
entered the mysterious place full of plants and potions.

"What shall I do?" he wondered. "Father is in great pain.
I've got to bring him something."

4

Radius looked at several bottles.

"Hmmm," he murmured. "This says 'Fire Belly.' Father has a fire in his belly. Maybe this will cure it."

Radius picked up the bottle and hurried back to Sir Cumference.

5

The knight gratefully took the bottle and gulped the liquid down. KABOOM!
Sir Cumference disappeared and a dragon now sat on the grass.

"AHHHHHH!" screamed Radius. "Where is my father?"

"H-h-here," hissed the dragon, flames slithering out of its mouth.

"H-h-help me," it pleaded, and belched puffs of smoke into the sky. "I f-f-feel beastly!" Its arms pedaled wildly as its tail thrashed back and forth.

"Don't worry, Father! I'll get help," Radius called over his shoulder as he ran back toward the castle.

The guards on watch had also seen the big explosion. When the smoke cleared, they saw the dragon and alerted everyone in the castle to the danger. Plans were made. Messengers were sent across the countryside to ask nearby knights to come and vanquish the fire-breathing beast.

9

"Mother, Mother!" called out Radius as he hurried into the castle.

"There you are!" answered his mother, Lady Di of Ameter, looking relieved. "Come inside quickly! A dragon has been spotted nearby."

"I know," Radius answered. "It's Father!" Radius told his mother what had happened.

"We must find another potion that will change your father back into himself. I'll go look for the doctor," she said. "We don't have much time. The knights plan to slay the dragon tomorrow morning."

Radius ran back to the doctor's
workroom. He looked at drawings and
notes. He peeked inside boxes and bags.
He searched through book after book.

Finally, he spotted a curious-looking
container with a set of spoons and a
poem. It might be the cure.

The Circle's Measure

Measure the middle and circle around,
Divide so a number can be found.
Every circle, great and small —
The number is the same for all.
It's also the dose, so be clever,
Or a dragon he will stay . . .
forever.

"Measure the middle and circle around…" thought Radius. "I'll bet Geo of Metry can help! Carpenters measure things every day."

Geo was with his brother Sym when Radius arrived at their workshop. They were looking at a wheel Sym had made.

"No matter where you look around the circle, the spokes go across the middle and cut the circle exactly in half," Sym said.

"Such a lovely design," complimented Geo.

When Radius heard "across the middle" and "around the circle," he had an idea. He needed to go back and reread the poem.

"Goodbye!" he called as he rushed out.

As he ran through the kitchen, he saw his cousin, Lady Fingers. She was baking pies.

She measured strips of dough into equal lengths with a span of her fingers. "Help me finish the last pie, please," she pleaded.

Radius thought about Sym's wheel. He arranged the strips on the pie like the spokes. There were three strips left over. He draped them around the rim of the pie pan.

"One, two, three strips go almost all the way around the edge. Pinky, may I have a little more dough?" asked Radius. Pinky was Lady Finger's nickname.

Lady Fingers handed him another strip of dough. Radius folded it in half, but half was longer than he needed. He folded it in quarters, but even a quarter of the piece was too long. He folded it in eighths, and an eighth was almost right.

Almost, but not exactly. "Pinky, I've got to get going," said Radius. "Farewell!"

Radius knew the magic number was more than three, but exactly how much more did it take to make a whole circle?

Back in the workshop, Radius found Geo's measuring tape — a long strip of cloth marked in inches. He measured 49 inches across the middle of a wheel and 154 inches around the outside edge. "154 divided by 49 is $3\frac{1}{7}$," he figured out.

Radius measured big wheels and small wheels. Every time, the distance around was $3\frac{1}{7}$ times the distance across. He stuffed the measuring tape into his pocket and ran off to the doctor's room.

As Radius ran around a corner, he saw his mother coming toward him. "There you are!" Lady Di cried. "I cannot find the doctor anywhere. What are we going to do?"

"It's alright, Mother. I know what to do." He showed her how he figured out the magic number.

He measured an onion slice, a basket, a bowl, and a round cheese.

Lady Di drew some diagrams to show the measurements. "This makes sense," she said. "Good work. I'll go and tell your father while you get the medicine. Don't forget we have only until morning."

about 11 inches

3½ inches

about 22 inches

7 inches

about 44 inches

14 inches

2½ inch wedge

17½ inches

55 inches

Radius went back to the doctor's workroom to get the medicine. "I think I understand these directions," he said out loud. "But if I measure the wrong dose, Father will remain a dragon forever!"

"I must be sure I am right," said Radius. He sat down to reread the poem. As he read, his eyelids slowly closed. Soon, he was fast asleep.

When Radius awoke, the sun was rising. "Oh, no, I must hurry," he said. He grabbed the medicine and headed toward the woods. The smoke and flames led him right to Sir Cumference the dragon.

The sleeping dragon was lying in a circle with its tail in its mouth. Just then, it opened one eye and hissed, "H-h-h-hello, S-s-son."

"Father, you're a circle, too, but such a big circle!" Radius looked at the dragon doubtfully. Could the distance around the dragon circle really be three and one seventh times the distance across it?

Just then, they heard the blaring of trumpets. "Hurry, Radius!" Lady Di said. "The knights are coming!"

"**M**other!" Radius said. "Hold this part of the tape while I measure across the middle of the dragon circle."

Sir Cumference cried out, "Why are you measuring my diameter?" As he spoke, his fiery breath burned the measuring tape so that Radius could barely read the numbers. It looked like seven feet, but was it?

Together, Radius and Lady Di measured 22 feet around the outside of the dragon circle. Sir Cumference whimpered, "Must you measure my circumference NOW?" Radius quickly divided 22 by 7. The circumference of the circle divided by the diameter was three and one seventh!

Now the knights were surrounding them with swords drawn.

"Radius! Lady Di! We've come to save you!" they shouted.

It was now or "dragon forever."

Radius gave the creature three and a seventh spoonfuls.

27

KABOOM-FAROOM! A big explosion shook the forest. When the smoke cleared, a hearty, human Sir Cumference stood in their midst.

Sir Cumference leaned over and hugged his son. "Thank you, Radius, but how did you do it?" Radius explained about the pie and the measurements.

Sir Cumference and the knights listened with amazement. Cheering, they swept Radius up on their shoulders and returned to the castle.

"Pies! Let's eat some pies!" shouted Sir Cumference. "Radius saved my life because of a pie."

At the celebration, Radius held up a pie and explained what he had discovered. "I found out that the outside edge of a circle, called the circumference, is three and one seventh times as long as the diameter, which is the measure across its middle. It's true for any circle."

"I say we honor this new discovery," said Sir Cumference. "From now on, pie with an e will be for eating. Pi without an e will be the name of this number for all things round!"

Geo and Sym stepped forward and gave Radius a present.

"This is a drawing compass. May it lead you to other great discoveries about circles!" they said.

And Radius had the seat of honor for the entire Pi Celebration which lasted three days, three hours, and twenty-four minutes.

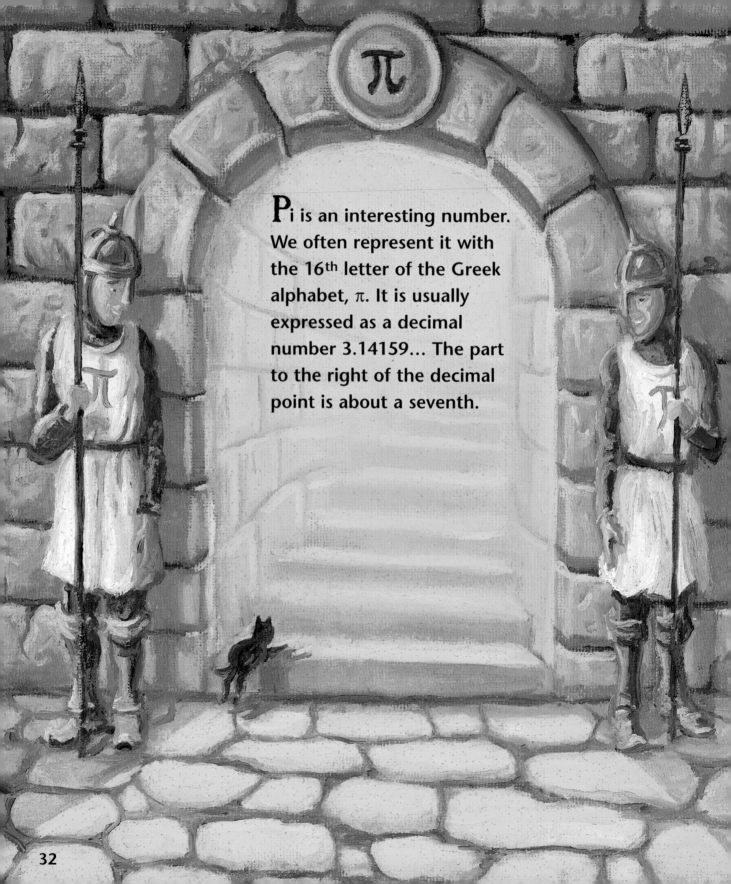

Pi is an interesting number. We often represent it with the 16th letter of the Greek alphabet, π. It is usually expressed as a decimal number 3.14159… The part to the right of the decimal point is about a seventh.